NATIONAL GEOGRAPHIC

Ladders

THE GOOD EARTH

SOIL

by Richard Easby

Where Food Begins

Farmers like Ismail Hassoun Hariri struggle to grow any crops in this parched land in Syria. Annual rainfall averages only 23 centimeters (9 inches). In some very dry years the barley crop fails to mature. Then it can only be used for feeding sheep and goats.

Syria's "Dead Cities"

Busy towns and villages once stood in this hilly region of northwestern Syria. Between the towns were large groves of olive trees. Centuries ago, this all changed. The people left these towns.

Today the land is treeless and rocky. Plants struggle to grow because the soil is thin and dry. It is hard to believe people ever lived here. What likely happened?

Syria's "dead cities" are ruins in a man-made desert.

The ancient farmers of this region grew and sold olives to make money. They cleared forests to plant olive trees. The forest canopy had protected the ground from sunlight and heavy rain. Now, the forest canopy was gone. The farmers had to protect the soil from **erosion** themselves. (Erosion happens when soil washes or blows away.) Centuries later, a war stopped the farmers from protecting the land. Soil slowly washed away. There was less soil to take in rainwater, so the land dried up. Underground springs stopped flowing because rain didn't soak into the ground. The people couldn't live here without good soil and water.

Today about seven billion people eat food grown in soil. Farmers need to produce more and more food. Scientists are worried that we are not taking good care of the land.

FACT   About 99.7 percent of human food comes from the land.

The Loess Plateau is important to farmers. Yet the land poses many challenges.

China's Loess Plateau

China's Loess Plateau formed over thousands of years. Wind blew silt into large piles of yellowish soil, called loess (pronounced "less"). The process that built up the loess is called **deposition.** The Loess Plateau is the biggest deposit of loess in the world. Many farmers have lived and farmed on this land.

The Loess Plateau is an example of how people can damage soil. In the 1950s, farmers started building terraces. Terraces are sections of flat ground carved into a hill. Terraces increase the amount of farmland. To build the terraces, farmers cleared all of the native plants growing on the steep slopes.

When dry, loess is as hard as rock. The terrace walls held up. The amount of crops increased for a while. But during each rainy season, the wet loess crumbled. Terraces fell apart. Without native plants, soil washed away. People couldn't rebuild the terraces fast enough, and the problem went on for decades. Eventually people had to leave or go hungry. The Loess Plateau became one of the fastest-eroding places on Earth.

Since then, the government has worked with farmers to slow erosion. Farmers plant trees and grasses that can hold soil in place, but erosion is still a problem. More must be done if farming is to succeed in this region.

FACT	China is losing its soil 40 percent faster than nature can replace it.

In Yunnan Province, China, things aren't so bad. Rice has been grown in paddies, or fields, like this one for centuries. Standing water shields soil from drying and erosion. Farmers often raise fish in the flooded fields. The fish manure adds nutrients to soil.

Scientists aren't sure whether Native Americans created the terra preta on purpose or by accident. But people recognized its value. A thousand years ago, two groups may have gone to war over control of terra preta.

Terra Preta in the Amazon

Soils in the Amazonian rain forest are fragile. If farmers cut down trees to grow crops, they expose the soil to heavy rains and very hot sun. The small amount of **minerals** and nutrients in the soil quickly washes away. The remaining soil becomes hard and dry. So scientists were surprised to find fertile soil along the Amazon River. Soil like this does not form in any tropical region. The soil was man-made.

The rich, black topsoil is called *terra preta*, meaning "black earth." How did the Native Americans who lived here make it? First, they burned plant matter at a low temperature to make charcoal. Next, they buried the charcoal in shallow holes along with food waste. The charcoal helped keep added nutrients from washing away. The quality of the soil greatly improved.

Eduardo Góes Neves (center, in blue shirt) is an archaeologist. He thinks ancient people created this soil by mixing charcoal with food waste.

Today people are using terra preta to improve soils in tropical regions around the world. Scientists hope it will lead to more food for the world's growing population.

FACT   It takes nature 500 years to replace 2.5 centimeters (1 inch) of topsoil.

7

The Sahel's Reclaimed Desert

A snapshot from 1986 shows that this place in Burkina Faso was a desert. Since then, villagers have re-greened the landscape.

The Sahel is the area between the Sahara and the humid forests of central Africa. The Sahel is hot and dry. In the 1950s, many people moved here and farming increased. Then two droughts occurred in the 1970s and 1980s. More than 100,000 people died from lack of food. The soil dried and hardened. When it finally rained, the ground couldn't take in the water. The water washed over the surface, taking bits of soil with it.

Farmers have slowly restored large areas of the Sahel. How did they do it?

Farmers placed long rows of stones across the ground. When it rained, water flowing over the dry ground was slowed by the stones. The soil was able to soak up the water. Water pulled fertile silt and plant seeds into the soil. The silt enriched the soil with nutrients. The seeds sprouted. The lines of stones became lines of plants. The plants slowed even more water. More plants grew, and the soil was further enriched. Within a few years, entire fields had been restored.

FACT About 850 million people are considered undernourished.

Issa Aminatou of Niger feeds her family with food she grows on once barren land. After the droughts, the United Nations Food and Agricultural Organization stepped in. It enlisted 10,000 women to plant millions of trees. Tree roots reduce erosion. The roots also help rain soak into the soil.

The Reed family lost a foot of soil from parts of their corn farm. To limit erosion, they changed how they farmed. Cletus Reed hopes his grandson will work these fields someday. "The land takes care of us as we care for it," he says.

Compacting the Great Plains

The Wisconsin Farm Technology Days is an annual show. Farmers go to check out the latest big farm machines, such as huge tractors and combines. Few people know that these big, heavy machines are crushing the life out of the land.

When a 15-ton combine rolls over a field, its weight compacts, or presses down on, the earth. Compaction is a danger to soil in the United States. Here farmers need huge machines to plant and harvest huge amounts of food.

Soil compaction is caused by heavy machines. It reduces harvests. The problem may cost U.S. farmers more than $100 million every year.

The rich soils in the Great Plains and the Midwest are made up of loose bits of sediment and minerals. These bits are held together with organic material. Tiny air pockets allow water and roots into the soil. Heavy machines squeeze out the air pockets. Roots can't grow and water can't soak in. Water flows over the compacted surface, causing erosion. Compaction and erosion both cause lower crop yields.

Today the U.S. produces more food than ever, but there's a problem. Modern farming practices are harming the soil. Some growers see this problem and are changing how they farm. They understand the need to care for soil so it will feed future generations.

FACT	Soil erosion costs the world $400 billion each year.

Check In	How is soil threatened? Why is it important to take care of soil?

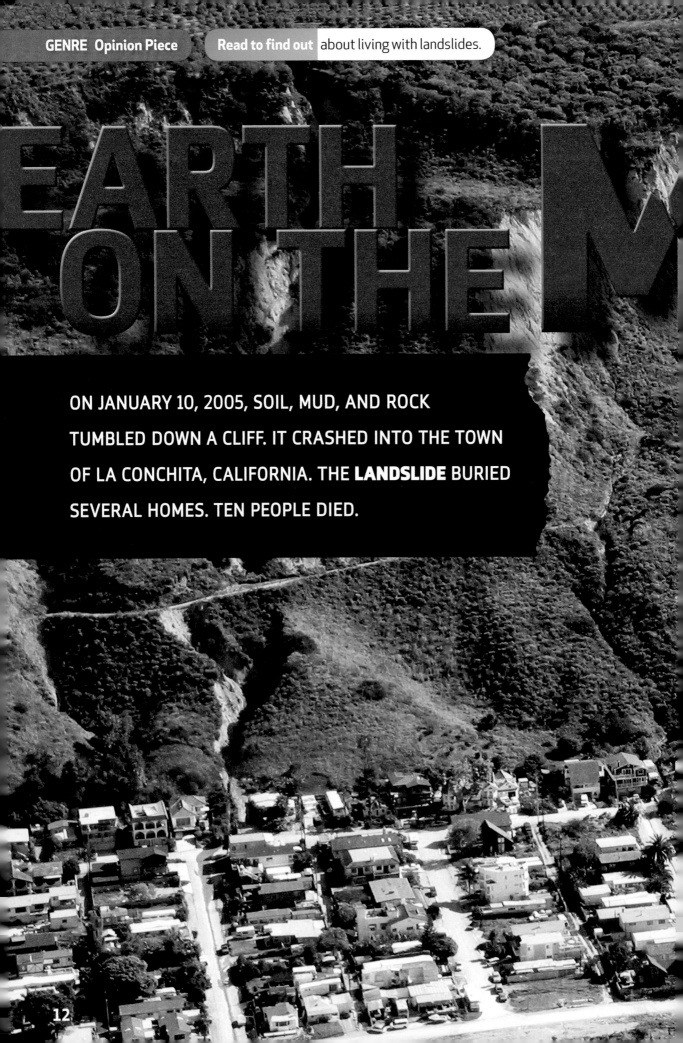

EARTH ON THE M

ON JANUARY 10, 2005, SOIL, MUD, AND ROCK TUMBLED DOWN A CLIFF. IT CRASHED INTO THE TOWN OF LA CONCHITA, CALIFORNIA. THE **LANDSLIDE** BURIED SEVERAL HOMES. TEN PEOPLE DIED.

MOVE

by Beth Geiger

This home in La Conchita was nearly buried by the 2005 landslide. Rescuers worked around the clock to free trapped residents.

Landslides happen all too often. Damage from landslides costs the United States over a billion dollars a year. Landslides ruin homes. They kill more than 5,000 people each year around the world.

Landslides often happen on slopes that have a layer of slippery clay under the ground. Heavy rains soak the soil. The slope becomes even more slippery. The pull of gravity does the rest. Earthquakes can also cause landslides. Landslides often race very fast. Others slip slowly over time. The slow movement of land down a slope is called creep.

Slow-moving landslides cause homes and roads to break apart. Fast-moving landslides can bury homes under tons of mud. Fast landslides happen without warning. The fastest landslides can move at speeds of up to 322 kilometers (200 miles) per hour!

Engineers try to stop landslides. They pipe extra water away so it can't soak into the soil. They also use concrete or steel pillars to help hold the slope in place.

Walls at the bottom of a slope can help hold back the soil. Homes on a slope can also be built to better withstand landslides.

What if a slope can't be made safe enough? The local government may ban people from building there.

In 2011, a slow-moving landslide destroyed several homes in Keene Valley, New York. Slow landslides are called creeps. This one was almost a mile wide. It was the largest recorded landslide in the state's history.

LANDSLIDES IN THE U.S. COST MORE LIVES EACH YEAR THAN ALL OTHER NATURAL DISASTERS COMBINED.

SHOULD PEOPLE BE ALLOWED TO LIVE WHERE A LANDSLIDE MIGHT STRIKE?

NO

Not all engineering measures work. For example, this stone retaining wall collapsed onto a road.

The government is responsible for keeping people from building in risky places. If a landslide strikes, government workers might be blamed for letting people live in an unsafe place. So engineers need to be sure a slope is safe before allowing people to build on it.

Engineering cannot stop all landslides. For example, take the Portuguese Bend landslide near Los Angeles, California. It is a slow-moving landslide. Engineering work was done to stop it, but the slope is still creeping downhill. The homes on it have been destroyed or damaged. Building on that slope is no longer allowed. One official wrote that this land will eventually slide into the ocean.

Taxpayers shouldn't have to pay for expensive fixes because some people won't move out of harm's way. In La Conchita, geologists say, the next landslide could happen at any time. Taxpayers may have to pay for engineering solutions. They may even have to pay to replace damaged homes.

YES

Workers use steel bars and concrete to build a retaining wall.
The wall will protect a new road from landslides.

There's risk everywhere. Lightning, falling trees, earthquakes, and other natural dangers cause damage. So why single out landslides? It is unlikely that your home would be damaged by a landslide.

Engineering can make some landslide zones safe. Look at Fidalgo Island, Washington. A coastal slope began to crack and move. Homeowners built a rock wall along the bottom. The wall helped keep waves from making the slope steeper. Engineers also pumped water away from the hill. The engineers say it's safe now.

People who are willing to take the risk shouldn't have to move. In La Conchita, people moved back in after a landslide in 1995 and another in 2005. They wanted to stay in their friendly town by the beach.

Check In   Which opinion do you agree with? What are your reasons?

DREAMS

by Beth Geiger

"The winds unleashed their fury with a force beyond my wildest imagination. It blew continuously for a hundred hours and it seemed as the whole surface of the earth would be blown away. As far as my eyes could see, my fields were completely bare." —LAWRENCE SVOBIDA

This farmer was describing the Dust Bowl. The Dust Bowl was a terrible disaster that affected the southern Great Plains in the 1930s. It caused millions of people to lose everything they had. It also caused people to change the way they farmed the land.

TO DUST

An Oklahoma farmer and his children head indoors during a dust storm.

The Dust Bowl began with a severe drought around 1931. Barely any rain fell in the southern Great Plains for the next eight years.

Wind carried away the topsoil needed for farming. Wind carried away the soil because fields were dry and bare. Black, sickening dust storms blocked the sun, turning day into night. Dust buried tractors and farmhouses.

The land was no longer rich and full of life. Something had gone very, very wrong.

Land of Promise

Before farmers arrived, the prairie was covered with prairie grass. The native plants were adapted to the prairie's dry conditions and strong winds. They could survive drought. Their roots helped hold soil in place.

Since the 1860s, the United States government had been giving away prairie land. People who wanted a farm traveled west to the prairie. These "homesteaders" were excited by promises of free land and rich soil.

Homesteaders wait to file land claims in Perry, Oklahoma, in 1893. A hand-painted sign above the door reads, "U.S. Land Office."

Homesteaders plowed under millions of acres of prairie grass. They planted wheat. At first, the rains fell and crops grew. During the 1920s, wheat prices soared and farmers made money. They bought tractors. They could plow even more prairie land and plant even more wheat. Times were good.

Hard work and the rich prairie soil made it all possible. No one worried about how they had changed the prairie by removing native plants and plowing the topsoil. They didn't think that the exposed soil might dry out and become as light as powder.

Tractor plows cut deeply into the soil.
This common practice led to drier, looser topsoil.
Fields were more vulnerable to strong winds.

"The soil is the one indestructible, immutable asset that the nation possesses. It is the one resource that cannot be exhausted; that cannot be used up."
—U.S. BUREAU OF SOILS, 1908

Starting Over

By the early 1930s, wheat prices dropped. Seeds were too expensive, so some farmers didn't plant. Their fields were left bare.

Then drought hit. Winds carried the powdery soil into the sky. This caused terrible dust storms that made life tough. People died of a new lung disease called "dust pneumonia." Animals starved to death. Farmers had barely enough money to feed their families.

On April 14, 1935, the worst dust storm of all happened. A black wall of dust about 2,438 meters (8,000 feet) high blew across Kansas, western Oklahoma, and Texas. That day became known as "Black Sunday."

A reporter named Ernie Pyle visited the Dust Bowl in 1936. He wrote, "It is the saddest land I have ever seen." Every month 10,000 people moved away from the southern Great Plains. They were looking for a new beginning.

Many others chose to stay. They owned their land and did not want to move to government camps. The government gave the farmers food. They also gave farmers money for the crops and animals that were lost. Finally, in 1939, the rains came. By then some parts of the prairie had lost 75 percent of the topsoil.

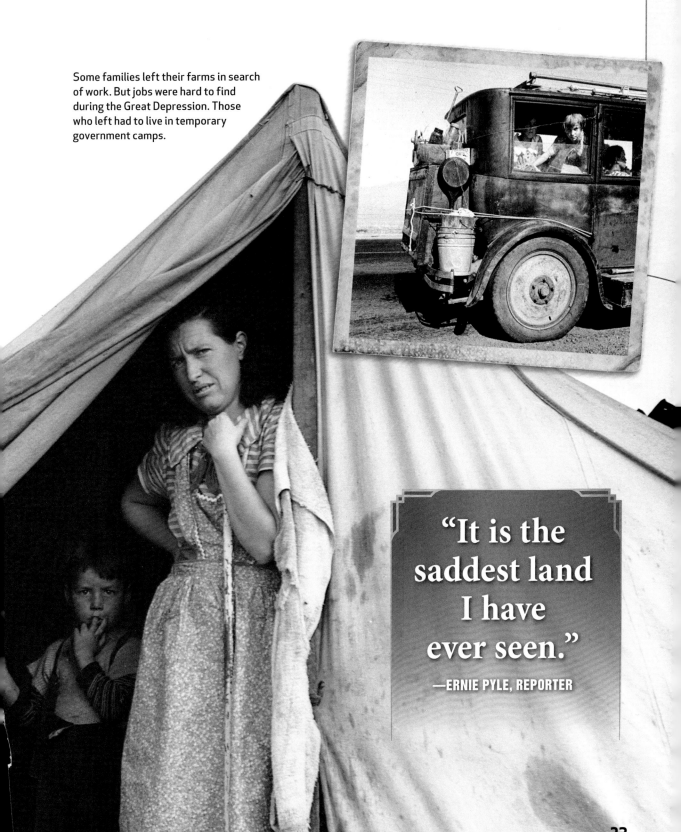

Some families left their farms in search of work. But jobs were hard to find during the Great Depression. Those who left had to live in temporary government camps.

"It is the saddest land I have ever seen."
—ERNIE PYLE, REPORTER

The Land Today

Hugh Hammond Bennett was a government soil scientist in Washington, D.C. He believed bad farming practices had caused the Dust Bowl.

Bennett wanted to change they way the Great Plains was farmed. He wanted to teach farmers how to protect the soil while farming.

On March 21, 1935, Bennett was trying to get the U.S. Senate to accept his soil conservation plan. Just then, dust from the Great Plains blackened the skies in Washington, D.C. The next day, President Roosevelt signed a law. This law helped start the U.S. Soil Conservation Service.

Dust from the Great Plains swept into Washington, D.C. Clouds of dust blocked the sun. Day turned into night.

Bennett's strategies to conserve soil are still used today. One strategy is contour farming. Farmers plant in curved in rows that follow the shape of a slope. Planting curved rows instead of straight ones helps prevent **erosion**. Another strategy is crop rotation. Farmers change which crops are grown on a field. This helps build soil nutrients.

Hugh Hammond Bennett (right), first Chief of the Soil Conservation Service

Bennett's strategies have helped farming continue in the southern Great Plains. But drought, erosion, and dust storms hit the area again in the 1950s, and still happen there today. Many say the Great Plains never fully recovered from those terrible dust bowl years.

In 2011, a major dust storm blew into Lubbock, Texas. Many people compared it to pictures they had seen of the Dust Bowl. Dust storms have become more common in the southern Great Plains because the region has been in a drought. If the drought continues, it could end up being one of the costliest natural disasters in U.S. history.

Check In What did people learn about soil from the Dust Bowl?

SAVING THE SOIL

by Beth Geiger

JERRY GLOVER says annual crops can be bad for the soil and bad for us. Glover is an **agroecologist.** He studies the link between food and soil. Glover grew up on a farm, but it wasn't until college that he realized how connected we are to soil. "Soil is where we get nutrients," he says.

Two wheat fields shimmer under the bright Kansas sky. One is like most wheat fields. The other may help change how food is grown.

Farming methods have improved since the Dust Bowl. But 80 percent of crops grown on Earth are **annuals.**

Food For Thought

Annual plants grow for just one year. Farmers must replant them from seed every spring. Annuals have shallow roots, too. Once the plants are harvested, they die. The bare soil is exposed to **erosion.** What soil needs is root systems and organisms that live in the roots all year. These things add nutrients. Without them, soil becomes less fertile. So farmers have to add expensive fertilizers. These fertilizers can harm the organisms that keep soil healthy. They can also get into the food we eat.

Jerry Glover helped create the field that might change how food is grown. The field is a research facility at the Land Institute. It is planted with grains that are **perennials,** not annuals. Perennials grow back year after year. Annuals do not. Perennials have deep roots that help enrich soil. No fertilization is needed. This saves money and helps keep nutrients in soil. Perennials also protect soil from erosion year round.

Growing, Growing, Gone

Why aren't more crops perennial? Annual grain crops usually produce more food than perennial grain crops. Jerry Glover plans to change this.

The part of most grains that's eaten is the seed. So growers have bred crops with big seeds for centuries. The crops produce high yields, or many seeds. But a lot of their energy goes into growing large seeds. The plants can't live through cold winters or droughts.

Perennial plants put a lot of their energy into surviving all year. That means growing deeper roots and larger leaves. They put less energy into growing the big seeds that farmers want.

Scientists bred this intermediate wheatgrass. It now grows bigger seeds, and more of them.

Wild intermediate wheatgrass has smaller seeds. Why? It puts more energy into growing roots that can survive winters and droughts.

Intermediate wheatgrass seeds

Annual wheat

Back to Roots

The challenge is to create new grain crops that are perennials and that produce many large seeds.

Most grain plants began in the wild as perennials. Then plant growers tamed them. Scientists at the Land Institute cross-breed the tamed annual crops with wild perennial crops. They choose perennial plants that have lots of larger seeds. These hybrid, or mixed, plants are new species. For example, a new species of perennial intermediate wheatgrass has been grown. The wheatgrass produces a new kind of flour called Kernza.

The goal of scientists is to create a new perennial crop plant. The plant would produce many large seeds and survive for many years.

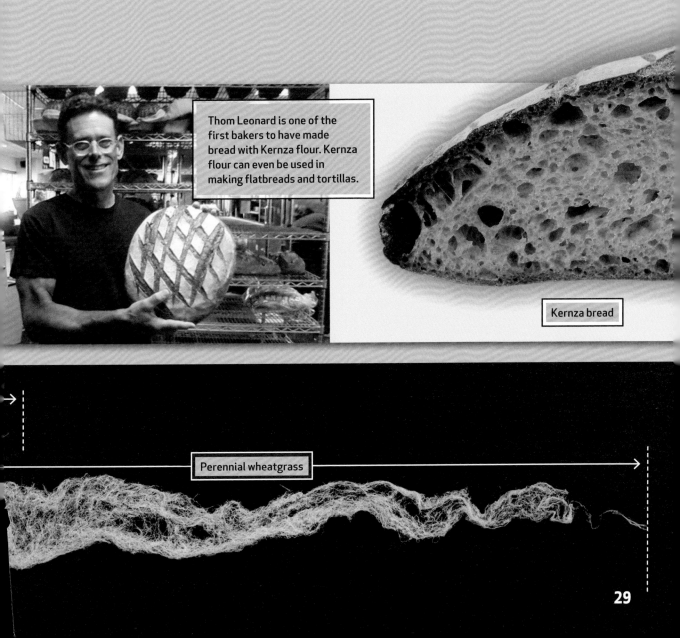

Thom Leonard is one of the first bakers to have made bread with Kernza flour. Kernza flour can even be used in making flatbreads and tortillas.

Kernza bread

Perennial wheatgrass

Rooting for Perennials

Jerry Glover sees perennial crops as the future of farming. He sees farm fields that won't damage soil or need lots of fertilizers. "We can think of agriculture in a whole new way," adds Sieg Snapp. Snapp is an agroecologist at Michigan State University.

Some perennial crops might not be ready for twenty years. Others might be ready sooner. Kernza cookies, anyone?

"Pigeon peas are pretty much the only crop which can be widely grown as a perennial grain," says Glover. The pigeon pea grows in tropical climates and has been shown to boost soil fertility. It's already a popular crop in Africa.

Illinois bundleflower has nutritious seeds. The seeds might be used to make cooking oil. So far, it's produced only small seeds and low yields. Snapp says Illinois bundleflower is "quite far" from being crop-ready.

Intermediate wheatgrass lasts only two years before it dies and has to be replanted. But scientists are improving it. One day its seeds might show up in your toast.

Sorghum is a warm-climate crop. It's grown for food and animal feed. In the U.S., perennial sorghum is "close to ready for prime time," says Snapp.

Check In   How might perennials help agriculture in the future?

Discuss

1. What do you think are some properties of soil that connect the four pieces in this book?

2. After reading "Earth on the Move," do you think people should be allowed to build houses on unstable land? Use examples from the text and your own ideas to support your answer.

3. What were some of the causes of the Dust Bowl? What were some lessons people learned from it?

4. Describe the problem Jerry Glover and others are trying to solve in "Saving the Soil." Then explain their solution.

5. What questions do you still have about soil? What research could you do to find out more?